TECHNICAL REPORT

A Funding Allocation Methodology for War Reserve Secondary Items

Kenneth J. Girardini, Carol E. Fan, Candice Miller

Prepared for the United States Army

Approved for public release; distribution unlimited

ARROYO CENTER

The research described in this report was sponsored by the United States Army under Contract No. W74V8H-06-C-0001.

Library of Congress Cataloging-in-Publication Data

Girardini, Ken.
   A funding allocation methodology for war reserve secondary items / Kenneth J. Girardini, Carol E. Fan, Candice Miller.
      p. cm.
   Includes bibliographical references.
   ISBN 978-0-8330-4934-6 (pbk. : alk. paper)
   1. United States. Army—Supplies and stores. 2. United States. Army—Appropriations and expenditures. 3. Resource allocation—United States—Planning. 4. Inventories—United States. 5. Military education—United States—Planning. 6. Military readiness—Planning. 7. Deployment (Strategy)—Planning. 8. Military planning—United States. I. Fan, Carol E. II. Miller, Candice. III. Title.

UC263.G54 2010
355.6'22—dc22

2010003440

The RAND Corporation is a nonprofit research organization providing objective analysis and effective solutions that address the challenges facing the public and private sectors around the world. RAND's publications do not necessarily reflect the opinions of its research clients and sponsors.

RAND® is a registered trademark.

Published 2010 by the RAND Corporation
1776 Main Street, P.O. Box 2138, Santa Monica, CA 90407-2138
1200 South Hayes Street, Arlington, VA 22202-5050
4570 Fifth Avenue, Suite 600, Pittsburgh, PA 15213-2665
RAND URL: http://www.rand.org/
To order RAND documents or to obtain additional information, contact
Distribution Services: Telephone: (310) 451-7002;
Fax: (310) 451-6915; Email: order@rand.org

Preface

U.S. Army units must be ready to deploy rapidly in the event of a contingency. During peacetime operations, the Army maintains inventory to support training and to maintain readiness. When a contingency occurs, deployed operating tempo often leads to increased demands for sustainment materiel for units involved in the operation, leading to an increase in global sustainment demands. Additional sustainment materiel is needed not only to maintain unit readiness in the face of these higher demand rates until the production base can respond but also to relieve the initial strain on the supply chain by reducing early airlift requirements. The war reserve secondary items (WRSI) portion of the sustainment stock within Army Prepositioned Stock (APS) is designed to address these two issues of production capacity gaps and early airlift requirements.

Historically, the computed WRSI requirements have not been fully funded. Yet no methodology exists by which war reserve requirements can be prioritized. Rather, after the requirements are computed, a time-intensive, decentralized review process is used to allocate resources to determine what portion of the requirement will be funded and where it will be positioned.

Thus, as part of an ongoing, formal process for determining WRSI stocks around the world, the Army asked the RAND Arroyo Center to develop techniques outside the Army's legacy system to prioritize item-level spending on war reserve materiel for a Northeast Asia contingency scenario with a known deployment schedule. The Army requested a quick-turn, 60-day product that (1) used empirical demand data to drive the allocation, (2) determined which items should be forward positioned versus stored in the continental United States (CONUS) and delivered via airlift, and (3) allocated the budgeted fiscal year (FY) 2007 funding.

This document should be of interest to logistics personnel, especially staff involved in inventory and stock positioning decisionmaking, and resource managers. This project was sponsored by the Assistant Deputy Chief of Staff, G-4, Headquarters, Department of the Army. The research was conducted within the RAND Arroyo Center's Military Logistics Program. RAND Arroyo Center, part of the RAND Corporation, is a federally funded research and development center (FFRDC) sponsored by the United States Army.

The Project Unique Identification Code (PUIC) for the project that produced this document is DALOC07572.

For more information on the RAND Arroyo Center, contact the Director of Operations (telephone 310-393-0411, extension 6419; FAX 310-451-6952; e-mail Marcy_Agmon@rand. org), or visit Arroyo's Web site at http://www.rand.org/ard/.

Contents

Figures

Tables

Summary

Army units must be ready to deploy rapidly in the event of a contingency, which creates challenges for the initial sustainment of deployed units, especially during the first 45 to 60 days or so it takes the first ships to arrive from CONUS. Additionally, when a contingency occurs, the increase in operating tempo leads to higher demands for some items. Because many items have lengthy procurement lead times, the baseline level of inventory will sometimes run out in the face of the higher demand before increased deliveries begin. Moreover, even if the contingency demands for certain items do not increase, these items must be transported to the site of the contingency, which places stress on the defense supply chain and on airlift capacity.

Stocks of war reserve secondary items (WRSI) within Army Prepositioned Stock (APS) are designed to address the two issues of production surge response time and competition for airlift early in a contingency. However, given the breadth of Army budget priorities, funding for WRSI stocks often falls short of the total calculated requirement, and the Army has lacked a formal method for prioritizing which items to stock.

Therefore, as part of an ongoing, formal process for determining WRSI stocks around the world, the Army asked the RAND Arroyo Center to develop techniques to prioritize the use of a $467 million FY 2007 budget for WRSI materiel for a Northeast Asia contingency scenario. RAND provided a quick-turn, 60-day product that (1) used empirical demand data to derive forecasts of the potential contingency demands, (2) determined which items should be forward positioned versus stored in CONUS and delivered via airlift, and (3) allocated the budgeted funding to maximize the WRSI inventory investment value with respect to readiness and reduced strategic airlift early in the contingency.

Empirical Demand Data Were Used to Forecast Contingency Demands

First, the Army needs a reliable method for estimating demands during a contingency. Second, it needs to know which items likely to be demanded are expected to have large demand increases over baseline levels, thus resulting in a shortfall or production gap. These items are important to be in WRSI inventory to bridge the gap until production surge deliveries commence to maintain readiness. Third, it needs to know which items would demand substantial early airlift if not stocked forward.

To determine which items fit the criteria above, RAND analyzed pre–Operation Iraqi Freedom (OIF) and OIF empirical data to identify items with large demand increases or with high backorders or stock availability problems early in OIF. Data from calendar year (CY) 2002 versus CY 2003 were used to identify the relative increase in demands at the beginning of OIF, as well as items with high backorders early in OIF. Items that experienced these problems reflect situations in which wartime demand stressed the production capacity. This process identified about 18,000 candidate items for possible war reserve funding.

RAND then used empirical data from ongoing contingency operations to develop two demand forecasts for a Northeast Asia scenario. We used demand data for units in OIF in CYs 2003 and 2006. CY 2006 data were used to ensure that parts for newly fielded and upgraded end items were included and that only items with ongoing demands were targeted as war reserve candidates. Demand data from both 2003 and 2006 were time-phased by unit to model the force buildup at the beginning of a contingency. Also, because the Army stocks materiel to support training and to maintain readiness for contingency operations, unless the item needs to be forward positioned, war reserve need be based only on the marginal increase in demands incurred as a result of contingency operations. So RAND determined the increases in global demands for these items, not just the contingency demands.

Forward Positioning Should Focus on Fast-Moving Items with Relatively Low Cost-to-Weight Ratios

Even if the supply system has sufficient inventory to handle increases in demand, decisions about which items should be forward positioned are key to ensuring that sustainment airlift is not overtaxed or supplies fall short during the initial days of a contingency. A forward-positioned item should be in high—and regular—demand, and the item should have a relatively low unit cost per pound or cubic foot. This enables most of the potential airlift need for sustainment, which is driven by item weight and size, to be avoided for relatively little inventory investment, which is driven by item price. For example, batteries are relatively heavy compared to their unit price. If they are needed in large numbers, it is much more cost-effective to forward position them than to use valuable airlift capacity to transport them from CONUS.

Only a small number of WRSI items need to be forward positioned to achieve a significant reduction in the airlift burden. Analysis using empirical demand data from OIF as a proxy for contingency demands identified about 1,800 candidate items for forward positioning. As shown in Table S.1, these items accounted for 70–80 percent of the volume (cubic feet) and about 80 percent of the weight of demands but less than 10 percent of the total demand value in the critical first 60 days of the demand forecasts, which were modeled from the March–April 2003 and time-phased January–February 2006 OIF demand streams. Thus, this policy requires relatively small investments in inventory to achieve very large airlift avoidance.

Table S.1
Forward-Positioned Items as a Percentage of All Items with Demands

Contingency Demand Data	Percentage of Items	Percentage of Demands	Percentage of Total Value of Demands	Percentage of Total Volume of Demands	Percentage of Total Weight of Demands
OIF, March–September, 2003	2	16	9	70	80
OIF, January–June 2006 (time-phased)	2	13	10	80	82

Items that are not forward positioned (referred to as *swing stocks*) can be used in any contingency, not just Northeast Asia. If the forward-positioned items are selected well, minimal airlift will be required to meet contingency demands using the remaining swing stocks (very expensive, low-demand, or small total shipment-weight items) stored in CONUS—e.g., one or two strategic lift aircraft per day early in a contingency.

Resource Allocation Method Trades Off Time and Readiness Benefits

RAND also developed a method to determine the best set of items (breadth) and inventory levels (depth) for a given funding level. This method takes into account the item's contribution to readiness and whether the item would be required early or late in the contingency—that is, how important it would be to potentially avoiding early airlift needs.

For this quick-turn analysis, RAND used the two demand forecasts developed from the OIF demand data (time-phased to reflect the deployment schedule assumed for the Northeast Asia scenario). However, demand forecasts from any source or model, including existing Army models, can be used as an input to the resource allocation tool. Using multiple forecasts reflecting different conditions allows the resource allocation to better deal with the real-world uncertainties associated with forecasting demands in a future contingency.

Another major advantage of this methodology is that the user may vary two levers—the weighting of the time periods (i.e., airlift avoidance) and the readiness weighting factor—to compute and compare different resource allocations that reflect different priorities. RAND used different weighting factors to develop two potential allocation schemes for the Army and then compared the benefits of stocking different items in terms of readiness and airlift avoidance. One of the solutions was chosen by the Army (after being reviewed and updated) as the basis for FY 2008 spending on WRSI materiel for a Northeast Asia scenario.[1]

Conclusion

Moving forward, the Army should ensure that the process for allocating war reserve budgets is flexible and agile so that it can be updated quickly as equipment, operational forecasts, and

---

[1] Due to changing priorities, the FY 2007 budget for WRSI materiel was shifted to other needs.

empirical demand data change. War reserve resources should be focused (1) on those items that should be forward positioned to avoid the excessive early sustainment burden and (2) on those items for which additional inventory minimizes the risk to operational readiness.

Acknowledgments

We would like to thank Thomas Edwards (Assistant Deputy Chief of Staff, Headquarters, Department of the Army, G-4 [Logistics]), who initially suggested the research effort based upon RAND Arroyo Center findings on the performance of the WRSI inventory in support of OIF. We would also like to thank then-LTG Ann E. Dunwoody (Deputy Commanding General, Army Materiel Command [AMC] and former Deputy Chief of Staff, G-4), LTG William E. Mortensen (former Deputy Commanding General, AMC), and MG Mitchell Stevenson (Commanding General, Combat Arms Support Command) for agreeing to and prioritizing this research effort at the meeting where the findings were reviewed and the research was suggested.

We are grateful to COL Kenneth Shreves, LTC William Kennedy, Vivian McBride-Davis, and Keith Mostofi at Headquarters, Department of the Army, G-4, for their assistance with this research. Additionally, they are putting it to use to improve the value of WRSI funding.

At AMC G-3, we would like to thank Kevin Maisel and his staff for assisting in the early formulation of the research.

Within RAND, Mahyar Amouzegar, Kip Miller, Eric Peltz, and Rick Eden provided valuable feedback on the structure and content of the document.

Abbreviations

ACOM — Army command
APS — Army Prepositioned Stock
ARCENT — Army Central Command
CONUS — continental United States
CTASC — Corps/Theater Automated Service Center
CY — calendar year
DLA — Defense Logistics Agency
DoD — Department of Defense
EDA — Equipment Downtime Analyzer
FedLog — Federal Logistics Catalog
FORSCOM — Forces Command
FY — fiscal year
HMMWV — high-mobility multi-purpose wheeled vehicle
LOGSA — Logistics Support Agency
NIIN — National Item Identification Number
OIF — Operation Iraqi Freedom
SARSS — Standard Army Retail Supply System
SSA — Supply Support Activity
SWA — Southwest Asia
TRADOC — U.S. Army Training and Doctrine Command
USAREUR — United States Army Europe Command
WRSI — war reserve secondary items

CHAPTER ONE

Introduction

Army units must be ready to deploy rapidly in the event of a contingency. For peacetime operations, the Army stocks materiel in support of training and to maintain readiness. When a contingency occurs, deployed operating tempo often leads to increased demands for sustainment materiel for units involved in the operation, leading to an increase in global sustainment demands. However, many items have lengthy procurement lead times, so the baseline level of inventory will sometimes run out in the face of the higher demand before increased deliveries begin. Moreover, contingency demands for sustainment materiel must be delivered to a different region of the world, placing stress on the defense supply chain and requiring sustainment airlift that must compete with other pressing airlift needs at the outset of a contingency.

The war reserve secondary items (WRSI) portion of the sustainment stock portion of Army Prepositioned Stock (APS) is designed to address these two issues of production capacity response time and competition for early airlift. In particular, WRSI inventory provides items[1] needed not only to maintain unit readiness in the face of potentially higher demand rates until the commercial and industrial base can respond, but also to relieve the initial strain on the supply chain by reducing early airlift requirements. War reserve materiel could be thought of as an early delivery to immediately increase global on-hand supplies, adding to the baseline level of stock, and to reduce early airlift requirements.

Historically, the computed WRSI requirements have not been fully funded, yet no formal methodology currently exists by which war reserve requirements can be prioritized. Rather, after requirements are computed, a time-intensive, decentralized review process is used to allocate resources and determine what portion of the requirement will be funded. Previous RAND analysis comparing Operation Iraqi Freedom (OIF) demand to WRSI requirements for a similar Southwest Asia scenario showed major discrepancies, largely due to the use of outdated input data in Army legacy systems. Additionally, the positioning of WRSI inventory—either forward or in the continental United States (CONUS)—had been found to be misaligned with actual sustainment airlift in contingencies and with the current principles being used by the Army to determine forward-positioned inventory.[2]

---

[1] WRSI includes items from Supply Classes 1 (subsistence), 2 (clothing), 3P (petroleum, oil, lubricants), 4 (construction materiel), and 9 (repair parts). In this analysis, we focused on items in Supply Classes 2, 3P, 4, and 9. For convenience, we will use the term "items" to refer to items in any of these supply classes.

[2] See Eric Peltz, Marc Robbins, Kenneth J. Girardini, and John Halliday, *Sustainment of Army Forces in Operation Iraqi Freedom: Major Findings and Recommendations*, Santa Monica, Calif.: RAND Corporation, MG-342-A, 2005, and Eric Peltz, Kenneth J. Girardini, Marc Robbins, and Patricia Boren, *Effectively Sustaining Forces Overseas While Minimizing Supply Chain Costs: Targeted Theater Inventory*, Santa Monica, Calif.: RAND Corporation, DB-524-A/DLA, 2008.

Therefore, as part of an ongoing, formal process for determining WRSI stocks around the world, the Army asked the RAND Arroyo Center to develop techniques outside the Army's legacy system and to prioritize item-level spending of $467 million allocated in fiscal year (FY) 2008 for war reserve materiel for a Northeast Asia contingency scenario with a known deployment schedule. The Army requested a quick-turn, 60-day product that (1) used empirical demand data to drive the allocation, (2) determined which items should be forward positioned versus stored in CONUS and delivered via airlift, and (3) allocated the budgeted FY 2007 funding.

This document describes the methodology and logic used to develop the resource allocation and forward positioning for that $467 million of war reserve sustainment. Chapter Two describes how empirical demands were used to reduce the scope of the allocation problem by limiting the population of items considered for this quick-turn effort. It also explains how items were identified for forward positioning. Chapter Three explains the resource allocation methodology, that is, how items were assigned relative benefits and compared to optimize the list of items to be purchased and forward positioned. The chapter also includes a description of how to base war reserve on the difference between peacetime training demands and forecasts of contingency demand—i.e., how to subtract or "offset" training demands from forecasts of contingency demand. Because inventory already exists in the military supply system to support training, war reserve should be based only on the marginal increase in demands as a result of contingency operations. In Chapter Four, we describe two potential solutions for prioritizing the use of FY 2007 budgeted WRSI funding and present conclusions. One of these solutions was the basis for FY 2008 spending on war reserve materiel for a Northeast Asia scenario.

Two appendixes are included. Appendix A describes the changes in demand that occurred with the onset of OIF and provides an example of the degree to which demands can increase during a contingency. Appendix B provides more detail on the resource allocation methodology used to prioritize items for war reserve resource allocation.

CHAPTER TWO

Identification of Candidate Items

The breadth of items required at the beginning of a contingency can be very large. For example, in the first six months of OIF, more than 100,000 different items were ordered by units in the theater. However, because of financial and operating environment constraints, not all items expected to be used during a contingency can be purchased for war reserve inventory.

For this quick-turn analysis, we limited the list of items for consideration by focusing on items needed to serve two main goals of WRSI inventory: (1) providing items that otherwise would likely be in short supply in a contingency and (2) "right-sizing" the sustainment airlift requirements at the beginning of a contingency. This chapter describes how historical demand data were used to produce the set of candidate items for possible war reserve allocation.

The Purpose of WRSI Inventory

First, for the purposes of supporting readiness by bridging any potential production capacity gap, those items with relatively long procurement lead times and large expected demand increases were included in the candidate list for war reserve. The amount of inventory potentially needed in war reserve depends on

- the procurement lead-time, which varies among items
- the expected demand increase, which varies depending on item failure modes and the deployed operating tempo.

For example, replacement of some items is almost directly proportional to their usage, whereas other items may be time-based, regardless of usage. Additionally, the relative contribution of items to keeping equipment operational varies and should be considered when selecting items for war reserve.

Second, aside from supporting readiness, how war reserve materiel affects supply-chain costs and sustainment airlift requirements is important as well. For example, batteries are relatively heavy compared to their unit price. If they are needed in large numbers, it is more cost-effective to forward position batteries than to use valuable, potentially scarce, airlift capacity early in a contingency to transport them to the theater. Thus, to minimize total supply chain costs while ensuring responsive, effective support, the Department of Defense (DoD) has started using a centralized theater inventory of high-demand, high-weight, low-cost items replenished by sealift. These items can then be cost-effectively delivered by ground convoy or intratheater air on demand. Achieving similar response times from CONUS would require more expensive intertheater airlift. This policy requires relatively small investments in inventory to achieve very large airlift savings.[1] Thus, items that are high-demand, high-weight or high-volume, and low-cost should also be included in the candidate list for war reserve.

In summary, based on the dual purposes of WRSI, the following items were included in the candidate list for war reserve:[2]

- items that are predicted to increase in demand relative to training demands
- items whose predicted increase in demands would stress the supply system
- items essential to operational readiness
- items that are high-volume or have low cost-to-weight ratios, i.e., relatively cheap to buy for inventory versus the cost of airlift required to move the materiel.

Because of the inherent difficulty in forecasting items that are predicted to increase in demand, we included the third category of items above for risk-mitigation purposes.

Criteria for Candidate Item Selection

The criteria for selection as a candidate item require knowing production lead times, expected demand increases, readiness contributions, and item-level volume, weight, and price information.

Lacking item-level production lead time and forecasted demand increase information for the quick-turn analysis requested by the Deputy Chief of Staff, G-4 (Logistics), we developed proxy measures for identifying those items for which the supply chain could not keep up with demand and/or for which demand rates increased: items with large demand increases or items with high backorders or stock availability problems early in OIF.[3] Items that experienced these problems reflect situations in which wartime demand stressed the production capacity.

We used item-level demand data for units in OIF from the Corps/Theater Automated Service Center (CTASC)[4] in calendar years (CYs) 2002, 2003, and 2006. CYs 2002 and 2003 were used to identify the relative increase in demands at the beginning of OIF, as well as items with high backorders early in OIF. CY 2006 was used to ensure that parts for newly fielded and upgraded end items were included and that only items with ongoing demands were targeted as war reserve candidates. Demand data from 2006 were time-phased by unit

---

[1] See Peltz et al. (2008). However, if a contingency occurs in a place where the United States does not have permanently stationed forces, a theater-level warehouse would not be in place and could take months to establish, involving both the decision and then the initial sealift deliveries. In the meantime, responsive support could require significant and what would be excessive sustainment airlift. Thus, a secondary role for WRSI inventory is to "initialize" this theater warehouse. In this situation, war reserve materiel could be on ships to be sailed quickly to a contingency region.

[2] The Army's WRSI process has always included some criteria to determine a subset of candidate items out of the total population of items. For example, in the past, essentiality code has been used.

[3] Although stochastic forecasts may be used in this methodology, the Army requested the use of empirical demands; historically, stochastic models have not always led to good results for the Army due to the difficulty in keeping information such as bills of material, failure rates, and expected demand increases updated for current National Item Identification Numbers (NIINs).

[4] CTASC document history files are compiled from data supplied by the Standard Army Retail Supply System (SARSS) computer. These data were obtained from the Logistics Support Agency (LOGSA) for CYs 2002, 2003, and 2006.

to model the force buildup at the beginning of a contingency. For example, the M1114 high-mobility multi-purpose wheeled vehicle (HMMWV) was fielded to units in Iraq after 2003; demands for items unique to the M1114 would only appear in the 2006 OIF demand data.

Table 2.1 shows five types of National Item Identification Numbers (NIINs) that were targeted for possible inclusion in war reserve, along with qualification criteria and examples of each.[5] The result of this empirical WRSI candidate selection process was that about 18,000 items qualified for potential stockage in war reserve.

Table 2.1
Five NIIN Attributes Targeted for Inclusion in War Reserve

NIIN Attribute	Qualification Criteria	Example
Experienced significant increase in demands between 2002 and 2003, with continuing demands in 2006	NIINs that were readiness drivers had to experience: a 25% increase in quantity demanded between CYs 2002 and 2003 at least 15 demands in CYs 2002 and 2003 at least 5 demands in CY 2006	Track for all tracked vehicles. Depending on vehicle type, demands for track increased 45–110 percent
Experienced moderate increase in demands between 2002 and 2003 and suffered significant backorders in 2003, with continuing demands in 2006	NIINs that were readiness drivers had to experience: a 10% increase in quantity demanded between CYs 2002 and 2003 at least 15 demands in CYs 2002 and 2003 at least 5 demands in CY 2006 at least 15 backorders in CY 2003	Brake shoes for rough terrain forklifts
Experienced significant demands in 2006	NIINs that were readiness drivers had to experience at least 25 demands in CY 2006, but no demands in CY 2002	Items for the M1114 HMMWV (which were not deployed in SWA in the initial deployment)
Essential to readiness with continuing demands in 2006 (even if they did not experience an increase in demands between 2002 and 2003) for risk-mitigation purposes	NIINs that are linked to readiness through deadlining reports had to experience continuing demands in 2006	M1 tank engine experienced only a 4 percent increase from 2002 to 2003 M88 engine experienced a 4 percent decrease in demand from 2002 to 2003
Should be forward positioned	NIINs with: unit cost- to-weight ratio less than $10/pound and total shipment weight greater than 1000 pounds or unit cost-to-volume ratio less than $20 per cubic feet and total shipment volume greater than 200 cubic feet	Storage batteries and tires for wheeled vehicles

---

[5] These thresholds were derived through experimentation and may require adjustment for different contingencies. To increase the emphasis on readiness drivers in war reserve, the qualification thresholds for the first three categories in Table 2.1 were more liberal for readiness drivers than those for non-readiness drivers. Readiness drivers were defined using the database underlying the Equipment Downtime Analyzer (EDA), a history of Army-wide equipment deadlining "O26" prints. See Eric Peltz, Marc Robbins, Patricia Boren, and Melvin Wolff, *Diagnosing the Army's Equipment Readiness: The Equipment Downtime Analyzer,* Santa Monica, Calif.: RAND Corporation, MR-1481-A, 2002.

Forward-Positioned NIINs

The fifth category shown in Table 2.1 consists of NIINs that should be forward positioned to provide initial stocks for a theater warehouse so as to avoid excessive sustainment airlift needs at the beginning of a contingency.

Again, lacking item-level forecasts of demand for a Northeast Asia contingency scenario, we used empirical demands from OIF as a proxy for forecasts of contingency demands to identify those items that should be forward-positioned. The analysis to identify forward-positioned NIINs used item-level CTASC demand data for units in OIF from CYs 2003 and 2006. We used empirical data from March–April 2003 to model the demands required at the beginning of a contingency. As with the candidate item selection, January–February 2006 was also used to ensure that parts for newly fielded and upgraded end items were included and that only items with ongoing demands were targeted for forward positioning. Demand data from 2006 were time-phased by unit to model the force buildup at the beginning of a contingency. Item-level volume, weight, and price information referenced the 2006 Federal Logistics Catalog (FedLog).

In total, about 1,800 candidate items were selected for forward positioning. As shown in Table 2.2, these 1,800 items accounted for 70–80 percent of the volume (cubic feet) and about 80 percent of the weight of demands in the March–April 2003 and time-phased January–February 2006 OIF demand streams, but less than 10 percent of the demand value. Thus, only a small subset of NIINs identified for war reserve must be forward positioned to achieve a significant reduction in the airlift burden, and these items represent a small percentage of the total value.

The significance of the forward-positioning decision cannot be overstated. Even if the supply system is resourced to handle the demand increase for an item, the assets may not be properly positioned. For items with a relatively low unit cost per pound or cubic foot, forward positioning of inventory is more cost-effective than using scarce or expensive strategic airlift to move relatively inexpensive items early in a contingency.

The war reserve forward-positioning decision is also important, because it results in the need for inventory that is dedicated to a specific theater, e.g., Northeast Asia, and, in essence, "locked in place" there. Inventory in another location, such as inventory in CONUS used to support training in CONUS, cannot be used as a flexible substitute for such forward-positioned materiel. In other words, even if the supply system is resourced to handle the demand increase for an item, if forward-positioning decisions are not made well and the assets are poorly positioned, it can lead to excessive demand for sustainment airlift early in a contingency.

On the other side of the issue, there is tremendous significance in an item not being chosen for forward positioning. Once funded, assets that are not forward positioned can be flexibly applied to any contingency, not just Northeast Asia; they are often referred to as *swing stocks*. If these assets are selected well, it will require minimal airlift—e.g., one or two strategic lift aircraft per day early in a contingency—to deliver these assets to the theater when needed. This suggests that the bulk of future WRSI resource dollars will be applied to swing stocks and that it will take very little airlift to deliver those swing stocks in a contingency.

Table 2.2
Forward-Positioned Items as a Percentage of All Items with Demands

Contingency Demand Data	Percentage of Items	Percentage of Demands	Percentage of Total Value of Demands	Percentage of Total Volume of Demands	Percentage of Total Weight of Demands
OIF, March–September, 2003	2	16	9	70	80
OIF, January–June 2006 (time-phased)	2	13	10	80	82

CHAPTER THREE

Resource Allocation Methodology

After the list of candidate items was selected, including the subset of forward-positioned NIINs, we turned to the resource allocation methodology that simultaneously determined the best set of NIINs (breadth) and inventory levels (depth) for a given resource level.

We begin with an outline of the methodology that will be described in this chapter. First, we selected two streams of contingency demand data (we used two streams of data to increase the robustness of our analysis for the FY 2007 budget allocation). For this quick-turn analysis, item-level historical demands in OIF served as contingency demand data for the deployment schedule assumed for this Northeast Asia scenario. Second, offsets were applied to each set of contingency demand data to capture the difference between training and contingency demands. Because the Army stocks materiel in support of training and to maintain readiness for peacetime operations, war reserve need only be based on the marginal increase in demands incurred as a result of contingency operations. Third, we then used the two streams of post-offset contingency demand data to generate benefit functions, i.e., functions that model the benefit of stocking an item and that take into account whether the item was required early or late in a contingency. The benefit functions were also adjusted to account for differences in contributions to readiness. For robustness, we combined the benefit functions for each NIIN from the two different streams of contingency demand data into a single benefit function. Finally, we compared the benefit functions of each NIIN across all NIINs and selected inventory levels to maximize the benefit of the available resources.

A major advantage of this resource allocation methodology is that the user may vary two levers that will be explained later—the weighting of the time periods and the readiness weighting factor—to rapidly compute and compare different resource allocations that reflect different priorities with respect to the degree of emphasis on fulfilling early versus late contingency demands and item criticality.

Need to Account for Inventory Held to Support Training

Before we begin our description of the resource allocation methodology, we note that, as directed by the DoD,[1] war reserve requirements should be based on the marginal increase in

---

[1] DoD Directive 3110.6, November 9, 2000, on War Reserve Materiel Policy states: "To minimize investment, war materiel requirements shall be offset by starter and swing stocks (including peacetime operating and training stocks) and, whenever possible, materiel available through industrial base programs, host-nation support agreements, bilateral military arrangements (e.g., Acquisition and Cross Servicing Agreements under 10 U.S.C. 2341-50, reference (c)), and commercial sources."

demands as a result of contingency operations and not on the entire contingency demand forecast. This is because baseline inventory to support training has already been accounted for in the military supply system.

The extent of the financial impact of accounting for inventory to support training demands can be approximated by the data in Table 3.1, which shows the change in demands between CYs 2002 and 2003 by major Army command (ACOM). Demands in Army Central Command (ARCENT) increased $4.6 billion, but demands in Forces Command (FORSCOM) decreased by $0.9 billion and demands in United States Army Europe Command (USAREUR) decreased by $0.3 billion. Thus, training demands represent a meaningful portion (about 25 percent) of contingency demands, and it is important to base war reserve requirements only on the difference between training demands and forecasts of contingency demands because inventory is already held in the system to support the former.

Table 3.1
Increase in Value of Demands Between
CYs 2002 and 2006

ACOM	Increase ($ billions)
ARCENT	4.6
FORSCOM	−0.9
USAREUR	−0.3

NOTE: See Table A.1 for a breakdown by subset of NIIN.

Contingency Demand Data

The first step in the resource allocation methodology was the selection of contingency demand data. For this quick-turn analysis, item-level forecasts of contingency demand were unavailable; as a substitute, we used empirical OIF demand data to model the demands at the beginning of a contingency. The analysis used CTASC demand data for units in OIF in March–September 2003 and January–June 2006. Again, demand data from 2006 were time-phased by unit to model the force buildup at the beginning of a contingency. Aside from being used to increase the robustness of the solution, two demand streams were included to capture demands for recently upgraded items and items from newly fielded equipment and to identify items that had been phased out.

Computation of the Training Demand Offset

The second step in the resource allocation methodology was to subtract or "offset" training demands from forecasts of contingency demand, because the Army already stocks materiel to support training and to maintain readiness for peacetime operations.

Lacking item-level information on demand increases due to a contingency, we compared demands before and after OIF and used this information to compute what we call the *offset percentage* and *training demand offset quantity*. The offset percentage is the proportion of contingency demands that would likely have occurred anyway as units trained, even if there had not been a contingency. The training demand offset quantity is the quantity demanded of an item that can be attributed to training. Thus, to subtract training demands from forecasts of contingency demand, we remove the training demand offset quantity from the contingency demand quantity to arrive at what we call the *post-offset contingency demand quantity* on which the war reserve allocation can be based. We call the removal of training demands from contingency demand data *applying the training demand offset.*

The offset percentage and training demand offset can be computed as follows:[2]

$$\text{Offset Percentage} = 1 - \frac{(\text{Demands in CY 2003} - \text{Demands in CY 2002})}{\text{OIF Demands in 2003}}$$

$$\text{Training Demand Offset Quantity} = (\text{Offset Percentage}) \times (\text{Contingency Demand Quantity}).$$

For example, if the quantity demanded of an item in CY 2002 was 80, and the quantity demanded in CY 2003 was 100, of which 40 was for operations in OIF, then the offset percentage for this item would be

$$\text{Offset Percentage} = 1 - \frac{(100 - 80)}{40} = 50\%.$$

That is, 50 percent of the contingency demands for this item would likely have occurred anyway just to support unit training.

If the contingency demand quantity for this item was 28, then

$$\text{Training Demand Offset Quantity} = 50\% \times 28 = 14.$$

Applying the training demand offset, we see that the post-offset contingency demand quantity would be 28 − 14 = 14. In other words, the quantity of 14, not 28, is the value on which the war reserve requirement should be based.

Application of the Training Demand Offset

The post-offset contingency demand quantity reflects the marginal increase in demands attributable to contingency operations. But the timing of contingency demands is also important: Filling demands early in a contingency is critical to unit readiness and to minimizing operational risk. Thus, the training demand offset must be applied in such a way as to preserve the timing of contingency demands.

In addition, as previously stated, one purpose of WRSI is to supply additional sustainment materiel at the beginning of a contingency until the industrial base can respond. WRSI

---

[2] The offset percentage was limited to values between 0 and 100 percent. For readiness drivers, the maximum offset percentage was set to 80 percent to avoid offsetting the entire demand forecast for these important items.

inventory to satisfy initial contingency demands may arrive via airlift (non–forward-positioned items) or come from forward-positioned theater-based inventory that is replenished via sealift.

Inventory that is not forward positioned, i.e., it is positioned in CONUS, can be used effectively and efficiently from the start of a contingency, because the use of airlift to replenish these items is cost-effective and not overly burdensome to strategic airlift. Therefore, for non–forward-positioned items, the training demand offset may be applied starting from the first day of the contingency to reflect the fact that baseline inventory to support training is immediately available to support contingency demands via airlift.

On the other hand, forward-positioned inventory is intended to serve as starter stock for theater inventory and to be replenished via sealift. (In other words, by definition, the baseline level of stock in CONUS to support training demands cannot substitute for inventory that should be in theater at the very start of a contingency.) Thus, for forward-positioned items, the training demand offset is applied *after* the first 60 days—or other assumed sealift replenishment time—to reflect the fact that baseline inventory to support training is available to support contingency demands once the sealift channel has been established.[3] These concepts are illustrated in Figure 3.1.

Figure 3.1
The Training Demand Offset Is Not Immediately Applied to Forward-Positioned Items

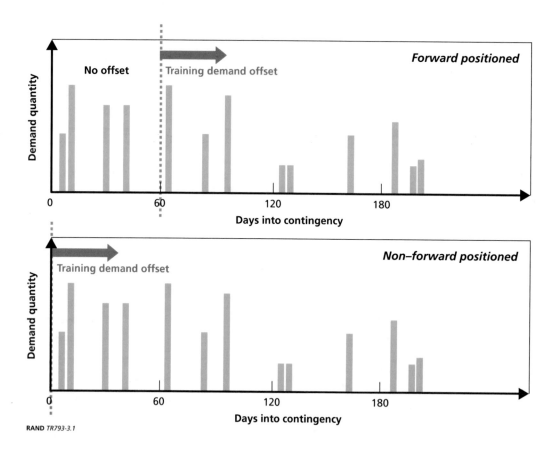

RAND *TR793-3.1*

---

[3] The Army is working with the Defense Logistics Agency (DLA) to set theater inventory levels to support training demands for units based in Northeast Asia. At a later date, this theater inventory may be integrated with WRSI, but this inventory was not accounted for in this quick-turn analysis.

To accomplish the dual purposes of preserving the timing of contingency demands and applying the training demand offset to reflect replenishment timelines, contingency demand data for each item were summed into ten-day time periods, or "buckets." The first bucket was the quantity of an item demanded in the first ten days of a contingency; the second bucket was the quantity of an item demanded in the second ten days of a contingency, and so forth. Because the contingency demand data streams each contained six months of data, there were 18 buckets for each item.

The training demand offset was applied to the buckets differently, depending on whether the item was intended to be forward positioned or not. For non–forward-positioned items, the training demand offset quantity was subtracted, in order, from the first bucket to the last bucket until the entire training demand offset quantity was subtracted from the contingency demand quantity.[4] For forward-positioned items, the training demand offset quantity was similarly subtracted, in order, from the seventh bucket (after 60 days or other assumed sealift replenishment time) to the last bucket. The result was post-offset contingency demand data grouped into buckets that preserved the timing of demands and that reflected replenishment timelines.

Subtracting the training demand offset in this manner meant that the remaining demands early in post-offset contingency demand data were for items that had no, or limited, training demand offset or that were intended to serve as starter stocks for theater inventory. *Thus, filling these early demands is critical to minimizing operational risk at the beginning of a conflict.*

Production Offsets

The training demand offset accounts for inventory already in the military supply system to support training and maintain readiness. Similarly, a production offset reflects additional inventory that the commercial and industrial bases can produce once they have had time to increase production capacity in response to higher demand rates.

A production offset should be applied to the contingency demand data after the time period required by the industrial base to respond to higher demand rates. This time period will typically vary by industry and commercial application, according to the item. In addition, most Army-managed items have lengthy procurement lead times that add to the time required by the industrial base to respond to increased demands, thus limiting the production offset potential very early in a conflict. Note that if the potential production capacity falls short of projected contingency demands, then war reserve would be needed to support forces for the duration of the conflict and would be replenished after the conflict ends.

Because of the quick turnaround nature of this project and the fact that there is limited production offset potential early in a conflict, a production offset was not applied to the contingency demand data. Army G-4 later asked the Defense Logistics Agency (DLA) to compute an item-level production offset based on the commercial and industrial base capacity.

---

[4]   See Appendix B for an example of the application of the training demand offset.

Construction of Benefit Functions

The third step in the resource allocation methodology was to construct benefit functions that modeled the benefit of stocking an item and that took into account whether the item was required early or late in a contingency.[5] The benefit functions were also intended to reflect each item's contribution to readiness.

At the end of the second step, we had constructed post-offset contingency demand data grouped into buckets that preserved the timing of demands and that reflected replenishment timelines. As previously stated, these values reflected demands for items that had no, or limited, training demand offset or that were intended to serve as starter stocks for theater inventory.

Using the average requisition demand quantity, we can relate the number of contingency requisitions that would be filled (benefit) to the amount of inventory needed to achieve this benefit in each ten-day time period, thereby still preserving the timing of the contingency demands. For example, if a contingency demand quantity of 6 in the first time period represented, on average, 1.5 requisitions, then a benefit of 1.5 would be achieved for an inventory level of 6 for this item.

The benefit in time periods after the first time period was then discounted to emphasize filling demands early versus late in a contingency. For example, if a contingency demand quantity of 9 in the second time period represented, on average, 2.25 requisitions, then an additional benefit of, say, $0.6 \acute{\mathrm{I}} 2.25 = 1.35$ would be achieved for an additional 9 in inventory, assuming that the discount factor was 0.6 in the second time period. The benefit function for this item is depicted in Figure 3.2.

Thus, the benefit of stocking an item was represented as a concave, piecewise linear function. Its graph consisted of line segments with positive, but decreasing, slopes, reflecting the fact that stocking increasing amounts of an item is of decreasing benefit.

For non-readiness drivers, the slope of each line segment depended on two factors: (1) the reciprocal of the average customer requisition quantity and (2) the weighting that is given to the time period, i.e., the *discount factor*. How long the slope was maintained was determined by the post-offset contingency demand quantity for the time period. If the item demand quantity was small for a time period, the slope was maintained for a smaller quantity.

We used the reciprocal of the average customer requisition quantity to model the benefit of filling contingency *requisitions*. In particular, the benefit was derived from the number of requisitions filled as opposed to the quantity filled.[6] Thus, we treated requisitions equally regardless of whether they were for items typically requested in small or large order quantities. For example, setting aside the weighting by the item's readiness contribution, filling one requisition for 100 washers was reflected in a total benefit of one requisition filled; filling one hundred requisitions for a quantity of one engine was reflected in a benefit of 100 fills. This method also allowed benefit to be given for partial fills.

---

[5]   More details about the construction of benefit functions can be found in Appendix B.

[6]   Customer requisitions are typically submitted by maintenance or operational personnel into the supporting Supply Support Activity (SSA). For spare parts, each requisition is assumed to be for one among several types of parts required to complete a work order. This assumption can be complicated by the existence of lower echelons of inventory that are replenished in lot-size quantities.

Figure 3.2
The Benefit Function of an Item Is Increasing at a
Decreasing Rate

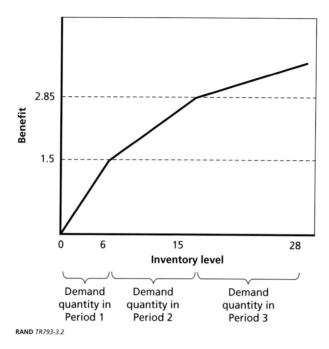

RAND *TR793-3.2*

The weighting of the time periods was determined by Army G-4, which indicated that the resource allocation should focus on filling the first 60 days of demand across both contingency demand forecasts. Therefore, more benefit was assigned to filling demands in the early time periods than to filling demands in later time periods. Subsequent time periods were discounted, or weighted less heavily, resulting in a concave function.

Readiness Drivers Receive Additional Benefit

For non-readiness drivers, the benefit was completely determined by the reciprocal of the average customer request quantity and the discount factor of the time periods. However, additional benefit was given to readiness drivers identified by the database of all deadlining demands used by the Equipment Downtime Analyzer (EDA).[7] That is, the benefit function for each item was multiplied by an empirically derived probability that the item would deadline an end item (the number of deadlining demands in the EDA database divided by the total number of demands). This concept is illustrated in Figure 3.3.

The benefit functions were also multiplied by a multiplicative *readiness weighting factor* chosen by the user. The larger the readiness weighting factor chosen by the user, the more benefit readiness drivers had compared to non-readiness drivers, therefore the more likely readiness drivers were selected for the solution.

---

[7] See Peltz et al. (2002). Our analysis used the percentage of demands in the EDA database versus the total demands Army-wide over the prior three years.

Figure 3.3
The Benefit Function of Readiness Drivers Is Increased
Based on How Critical the Item Is to Readiness

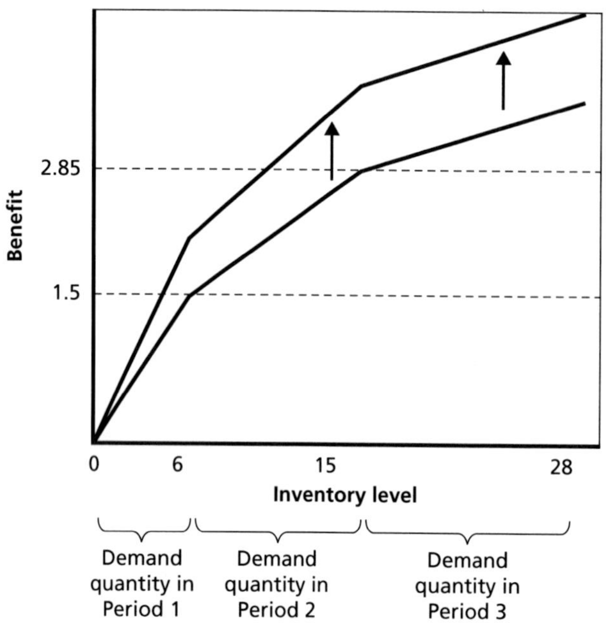

RAND TR793-3.3

Combining Benefit Functions Creates a More Robust Solution

After the benefit functions were created for each post-offset contingency demand stream, the two benefit functions were combined to create a single concave, piecewise linear benefit function as seen in Figure 3.4.[8] Combining multiple benefit functions increases the robustness of the solution by weighting items and quantities that are common across different contingency demand streams more heavily and avoids creating a solution that is tailored too closely to one specific instance of contingency demands.[9] Any number of alternative contingency demand streams can be linearly combined to improve robustness.

Comparison Across Benefit Functions

The final step in the resource allocation methodology was to compare the benefit functions across all items and use marginal analysis to maximize the benefit of the available resources. In short, we maximized the discounted number of contingency requisitions filled, constrained by a fixed budget.

The benefit-to-cost ratio for adding another asset of each item was computed by dividing the marginal benefit, i.e., the slope, by the unit price. These benefit-to-cost ratios were then

---

[8]   A linear combination of concave piecewise linear function is also concave, piecewise, and linear.

[9]   It is possible to give more weight to one contingency demand stream than to another. We chose to give equal weight to both demand streams because there was no reason to believe that the 2003 contingency demand stream was better than that from 2006.

Figure 3.4
Benefit Functions Can Be Combined to Increase the
Robustness of the Solution

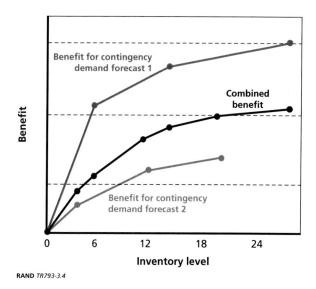

RAND *TR793-3.4*

ranked from highest to lowest. The inventory levels associated with the highest benefit-to-cost ratios were selected until the available budget limit was reached.[10] By investing the next dollar on the item with the largest benefit-to-cost ratio, limited funds were optimized to select the items providing the most benefit.

---

[10] Marginal analysis is optimal for this resource allocation. See Hugh Everett, III, "Generalized Lagrange Multiplier Method for Solving Problems of Optimum Allocation of Resources," *Operations Research,* Vol. 11, 1963, pp. 399–417; Bennett Fox, "Discrete Optimization via Marginal Analysis," *Management Science,* Vol. 13, No. 3, 1966, pp. 210–216; and Craig Sherbrooke, *Optimal Inventory Modeling of Systems: Multi-Echelon Techniques,* 2nd ed., Norwell, Mass.: Kluwer Academic Publishers, 2004. An example of marginal analysis can be found in Appendix B.

CHAPTER FOUR

Solutions and Conclusions

In this chapter, we present two solutions for the budgeted FY 2007 WRSI funding using different discount factors, i.e., weightings of the time periods, but the same readiness weighting factor. Then we present some conclusions and recommendations about war reserve requirements and resourcing.

Two Solutions

Two solutions were presented to Army G-4 for their consideration, using different weightings of the time periods but the same readiness weighting factor. For Solution 1, the discount factor for the first time period was 1; that for the second time period was 0.6; that for the third time period was $(0.6)^2$; and so on. In general, the discount factor for the nth time period was $(0.6)^{(n-1)}$.[1]

For Solution 2, the discount factors ranged from 1,000 for the first time period to 0.01 for the last time period, as seen in Table 4.1. Solution 2 was the result of weighting the first six time periods very heavily so that there was significantly more benefit to fulfilling contingency demands in the first 60 days than doing so later. Army G-4 chose Solution 2 to emphasize filling early forecasted demands and subsequently used it as the basis for FY 2008 spending on war reserve materiel for a Northeast Asia scenario.

Tables 4.2 and 4.3 show a comparison of the aggregate characteristics of the two solutions. Table 4.2 shows that the two solutions are very similar in terms of number of items and

Table 4.1
Discount Factors Used in Solutions 1 and 2

Solution	Time Period										
	1	2	3	4	5	6	7	8	9	10	10 + n
1	1	0.6	0.36	0.22	0.13	0.08	0.05	0.03	0.02	0.01	$0.6^{(n+9)}$
2	1,000	500	250.00	125.00	62.50	31.25	0.5	0.35	0.245	0.17	$0.17 \times (0.7)^n$

---

[1] The value 0.6 was chosen to exponentially decrease the weighting on successive time periods. Other values between 0 and 1 can also be used to achieve a concave, piecewise, linear benefit function.

Table 4.2
A Comparison of the Value, Volume, and Weight of the Solutions

Solution	No. of Items	Value ($ millions)	Volume (millions of cu. ft.)	Weight (millions of lb.)
1	15,347	470	3.86	75.1
2	15,042	468	3.09	60.6

NOTE: Although the resource constraint in the allocation was set to $467 million, the total resources used were slightly more due to rounding.

Table 4.3
A Comparison of the Item Characteristics of the Solutions

Solution	No. of Forward-Positioned Items	No. of Class 9 Items	No. of Army-Managed Items	No. of Non–Army Managed Items	No. of EDA Critical Items
1	1,679	12,563	2,179	13,168	9,066
2	1,682	12,333	2,072	12,970	8,926

dollar value, but Solution 1 produces more volume and weight than Solution 2. The item characteristics of the solutions are also very similar, as seen in Table 4.3.

Although the two solutions are relatively similar, the different discount factors resulted in solutions that were very different in terms of demand weight accounted for at the beginning of a contingency. Specifically, Solution 2 accounts for more of the early demand weight of the March–September 2003 and January–June 2006 OIF data.

Table 4.4 shows the total weight accounted for with respect to the March–September 2003 OIF data. Table 4.5 shows the same for the time-phased January–June 2006 OIF data. The weight accounted for was split into 10-day time periods. Only the first nine 10-day time periods of the contingency demands were included in the tables, because of space considerations, but these early time periods illustrate the difference between the two solutions. The third row in each table shows the difference in the total weight accounted for by Solution 1 versus Solution 2; it is the difference between the first and second rows. Negative values in this row are highlighted and indicate that Solution 2 accounted for more of the weight than Solution 1. For reference, the last row in each table is the total weight represented by the contingency demand stream. Comparing the first two rows to the last row, we see that both solutions account for a large proportion of the weight of the contingency demand streams.

From these tables, we can conclude that, as a result of the different time period weighting, Solution 2 allocated funding to more completely fill the early (first 60 days) demands in both forecasts. Solution 2 also placed additional emphasis on items that were to be forward positioned because these items did not have a training offset in the first 60 days.

Table 4.4
Total Weight (in Millions of Pounds) Accounted for by Solution, by Time Period, for March–September 2003 OIF Data

| Solution | Time Period | | | | | | | | |
	1	2	3	4	5	6	7	8	9
1	3.49	5.30	4.02	4.93	4.46	6.92	3.97	4.19	3.99
2	3.51	5.32	4.04	5.03	4.53	7.00	2.87	2.52	2.37
Solution 1 minus Solution 2	−0.02	−0.02	−0.02	−0.10	−0.07	−0.08	1.10	1.67	1.62
Total Weight in Demand Forecast	3.64	5.56	4.28	5.86	5.48	7.18	5.37	5.40	7.74

Table 4.5
Total Weight (in Millions of Pounds) Accounted for by Solution, by Time Period, for Time-Phased January–June 2006 OIF Data

| Solution | Time Period | | | | | | | | |
	1	2	3	4	5	6	7	8	9
1	0.95	2.25	3.99	5.65	8.58	4.06	3.78	4.69	3.71
2	0.95	2.26	4.01	5.74	9.27	4.28	2.89	2.70	2.54
Solution 1 minus Solution 2	0	−0.01	−0.02	−0.09	−0.69	−0.22	0.89	1.99	1.17
Total Weight in Demand Forecast	0.98	2.32	4.13	6.17	10.20	4.46	4.27	5.29	4.73

Conclusions and Recommendations

War reserve materiel is intended to relieve the initial strain on the supply chain by providing additional inventory to meet the increased operating tempo and demand rates of units in contingency operations until the production base can surge. In addition, forward positioning of war reserve should be used to avoid excessive early sustainment airlift requirements.

Based on the research completed for this analysis, we make the following recommendations for improving the war reserve requirements determination process:

1. Ensure that the requirements determination process is flexible and agile so that it can be updated quickly as equipment, operational forecasts, and empirical demand data change.
2. Base the requirements determination process on the increase in forecasted demands. That is, take into account inventory already in the supply system to support training.

3.   Focus war reserve resources on those items that should be forward positioned in order to avoid the excessive early sustainment airlift burden and on those items for which additional inventory minimizes the risk to operational readiness.

The first point is important, because the war reserve requirement determination process must be repeated each time war reserve funding is available, in order to update the benefit functions to reflect existing assets already in war reserve, to ensure that new and upgraded items are included in war reserve, and to ensure that outdated items are removed from war reserve. Attention must also be paid to items that are not used in peacetime, such as chemical protective suits.

The second point reemphasizes that the training demand offset can provide a significant resource to offset war reserve requirements and that this offset varies by item. A production offset should also be applied to reduce the war reserve requirement.

The last point—that the war reserve should focus on items that reduce operational risk and/or airlift requirements—serves to emphasize the goals of WRSI. Forward-positioned items should include low-cost volume and weight drivers to reduce the airlift requirement. Items that experience demand increases in a contingency, have long production lead times, and/or tend to experience supply chain shortages early in an operation—particularly those that are EDA critical—should also be included in war reserve to reduce operational risk.

In the near term, as demands from the units deployed in Southwest Asia (SWA) decrease, a potential opportunity exists that may allow the Army to benefit in two ways: (1) the increased supply chain inventory levels resulting from the earlier increased operating tempo may be used to fill the WRSI requirement as the operating tempo decreases, and (2) retaining the items required for forward positioned war reserve in SWA would allow inventory to be available in the theater. In both cases, the computation of appropriate inventory levels below which inventory should not be drawn down, i.e., the war reserve requirement, would enable the Army to avoid later unnecessary and expensive additional procurement of materiel to resource war reserve for SWA.

Changes in Demands with the Onset of Operation Iraqi Freedom

This appendix describes in more detail the changes in demand that occurred with the onset of OIF. Increases as well as decreases in demand accompanied the start of OIF. Table A.1 includes only NIINs that experienced an *increase* in demands between CYs 2002 and 2003, whereas Table A.2 includes only those that experienced a *decrease* in demands between CYs 2002 and 2003.

A comparison of Tables A.1 and A.2 shows that for the subset of NIINs for which CY 2003 demands were greater than CY 2002, there was an increase of $4.9 billion in demand value. For the subset of NIINs with a decrease from CY 2002 to CY 2003, there was a decrease of $1.4 billion in demand value. Table A.3 shows the change in demand by ACOM. The number of SSAs indicates the number of units supported by that ACOM. ARCENT

Table A.1
NIINs with an Increase in Demands Between CYs 2002 and 2003

Subset of NIINs	No. of NIINs (thousands)	Difference in Number of Demands in CY 2003 over Number of Demands in CY 2002 ($ millions)	Difference in Value of Demands in CY 2003 over Value of Demands in CY 2002 ($ billions)	Difference in Volume of Demands in CY 2003 over Volume of Demands in CY 2002 (100,000 of cu. ft.)
NIINs with demands in both CYs 2002 and 2003	109	3.0	4.5	60
NIINs with demands in CYs 2002 or 2003	185	3.2	4.9	62

Table A.2
NIINs with a Decrease in Demands Between CYs 2002 and 2003

Subset of NIINs	No. of NIINs (thousands)	Difference in Number of Demands in CY 2003 over Number of Demands in CY 2002 ($ millions)	Difference in Value of Demands in CY 2003 over Value of Demands in CY 2002 ($ billions)	Difference in Volume of Demands in CY 2003 over Volume of Demands in CY 2002 (100,000 of cu. ft.)
NIINs with demands in both CYs 2002 and 2003	68	−0.3	−1.1	−2.2
NIINs with demands in CYs 2002 or 2003	120	−0.4	−1.4	−2.6

Table A.3
Change in Demands Between CYs 2002 and 2003, by Army Command

ACOM	No. of SSAs in 2003	Difference in Number of Demands in CY 2003 over Number of Demands in CY 2002 ($ millions)	Difference in Value of Demands in CY 2003 over Value of Demands in CY 2002 ($ billions)	Difference in Volume of Demands in CY 2003 over Volume of Demands in CY 2002 (100,000 of cu. ft.)
ARCENT	107	3.6	4.6	20.0
FORSCOM	144	−0.6	−0.9	0.1
USAREUR	56	−0.4	−0.3	0.4
TRADOC	16	0.1	0.2	0.3

NOTE: Not all major commands were included in this table.

and the U.S. Army Training and Doctrine Command (TRADOC) incurred the bulk of the increase in demands, while FORSCOM and USAREUR were responsible for the majority of the decrease. In summary, the extent of the training demand offset is quite large and can represent meaningful reductions in war reserve requirements.

Details of the Resource Allocation Methodology

This appendix describes the details of the resource allocation methodology. First, we grouped or "bucketed" the contingency demand data into ten-day time periods to preserve the timing of demands. Second, we computed the training demand offset and applied it to the bucketed contingency demand data so that war reserve would be based on the marginal increase in demands due to the contingency. The application of the offset preserved the timing of the demands and reflected replenishment timelines. Third, using the post-offset contingency demand data from the second step, we constructed benefit functions that modeled the benefit of stocking an item. Finally, we compared the benefit functions across all NIINs to optimize the resource allocation.

Grouping of Contingency Demand Data

First, we grouped the contingency demand data to preserve the timing of the demands. Demands were grouped into ten-day time periods or "buckets." This translated into 18 ten-day buckets for the two six-month contingency demand streams of March–September 2003 and time-phased January–June 2006 OIF data.

The number of demands (in this case one demanded on each of seven days for different quantities)[1] and the sum of the total quantity demanded in each bucket was recorded. The sum of the quantity demanded or "bucket quantity" for each bucket was then aggregated. The cumulative bucket quantity represented the inventory level required to cover demands in the first 10, 20, 30, etc. days of the contingency demand forecast. These values were recorded in a contingency demand bucket table as shown in Figure B.1.

In the upper half of this figure, the timing and quantity of demands for an item are shown. In all, there are seven demands for the item with a total demand quantity of 28. Grouping the demands into ten-day buckets yields the cumulative bucket quantities 6, 15, and 28.

Application of the Training Demand Offset

Second, after the contingency demand data were bucketed to preserve the timing of the demands, we applied the training demand offset, that is, we subtracted training demands from the contingency demands.

---

[1] There can be multiple demands on the same day.

Figure B.1
Grouping of Contingency Demand Data

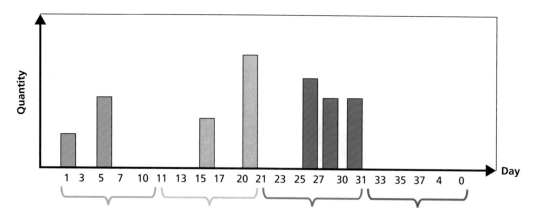

Assuming these are all the
demands for this NIIN, then:

Total number of demands = 7

Total quantity = 28

Contingency Demand Bucket Table

Bucket Number	Bucket Quantity	Cumulative Bucket Quantity
1	6	6
2	9	15
3	13	28

Cumulative Bucket Quantity represents inventory level
needed to cover demands in the first 10, 20, 30 days

RAND *TR793-B.1*

To do this, we first computed the training demand offset. In the example used in the main text, the quantity demanded of an item in CY 2002 was 80; the quantity demanded in CY 2003 was 100, of which 40 was for operations in OIF; and the offset percentage for this item was[2]

$$\text{Offset Percentage} = 1 - \frac{100 - 80}{40} = 50\%.$$

If the contingency demand quantity for this item was 28, then

$$\text{Training Demand Offset Quantity} = 50\% \times 28 = 14.$$

As previously stated, for non–forward-positioned NIINs, the training demand offset was subtracted beginning with the first bucket to the last bucket until the entire training demand offset quantity was subtracted from the contingency demand quantity. For forward-positioned NIINs, the training demand offset quantity was similarly subtracted in order from the seventh bucket (after 60 days or after other assumed sealift replenishment time) to the last bucket.

Figure B.2 shows the original and the resulting bucket tables—depending on whether the NIIN is forward positioned or not. In this example, the training demand offset quantity

---

[2] As previously mentioned, the offset percentage was limited to between 0 and 100 percent. For readiness drivers, the maximum offset percentage was set to 80 percent to avoid offsetting the entire demand forecast for these important items.

Figure B.2
The Application of the Training Demand Offset Changes the Contingency Demand Bucket

Original Bucket Table

Bucket Number	Bucket Quantity	Cumulative Bucket Quantity
1	6	6
2	9	15
3	13	28

Example: Offset percentage = 1 – (CY 2003 – CY 2002)/(OIF 2003) = 1 – (100-80)/40 = 50%

Training demand offset = 60% × (contingency demand forecast) = 50% × 28 = 14

Offset is applied to forward-stocked NIINs after first 60 days (beginning with Bucket #7)

Non–Forward-Stocked NIIN Bucket Table

Bucket Number	Bucket Quantity	Cumulative Bucket Quantity
1	0	0
2	1	1
3	13	14

Forward-Stocked NIIN Bucket Table

Bucket Number	Bucket Quantity	Cumulative Bucket Quantity
1	6	6
2	9	15
3	13	28

Since there are no demands after the first 60 days, no offset needs to be removed.

RAND *TR793-B.2*

of 14 was subtracted from the first bucket and part of the second bucket for the non–forward-positioned NIIN. There is no change in the bucket table for the forward-positioned NIIN, because there are no demands in the seventh to the last buckets.[3]

Construction of Benefit Functions

Third, after the post-offset contingency demand bucket tables were created, we constructed the benefit functions. Using the average requisition demand quantity, we related the number of contingency requisitions that would be filled (benefit) to the amount of inventory needed to achieve this benefit in each bucket.[4]

The benefit of stocking an item was represented as a concave, piecewise linear function, that is, as line segments with positive, decreasing slopes. For non-readiness drivers, the slope of each line segment depended on two factors:

- the reciprocal of the average customer requisition quantity
- the weighting that is given to the time period or bucket.

---

[3] Typically, there will be demands in Days 61–180, but for this example to fit we only showed demands through Day 30.

[4] Customer requisitions are typically submitted by maintenance or operational personnel into the supporting SSA. For spare parts, each requisition is assumed to be for one among several types of parts required to complete a work order. This assumption can be complicated by the existence of lower echelons of inventory that are replenished in lot-size quantities.

Specifically, the slope of each line segment was given by the equation

$$\text{Slope} = \frac{\text{Total Demands}}{\text{Total Quantity}} \times \text{Discount},$$

where the bucket number determined the discount factor.

How long the slope was maintained was determined by the post-offset contingency demand quantity for the period.

For example, in Figures B.3 to B.5, we continue with the previous example and construct the benefit function for a forward-positioned item. Assume that the discount of the nth bucket was equal to

$$\text{Discount for } n\text{th Bucket} = (0.6)^{(n-1)},$$

then the slope of the line segment, which represents the benefit, would be:

$$\text{Slope} = \frac{7}{28} \times (0.6)^{(1-1)} = 0.25 \,.$$

This slope would be maintained for six units, the quantity in the first bucket.

The slope of the line segment, or benefit, in the second ten-day time period would be:

$$\text{Slope} = \frac{7}{28} \times (0.6)^{(2-1)} = 0.15 \,.$$

This slope would be maintained for nine units, i.e., between inventory levels 7 and 15.

The slope of the line segment, or benefit, in the third ten-day time period would be

$$\text{Slope} = \frac{7}{28} \times (0.6)^{(3-1)} = 0.09 \,.$$

This slope would be maintained for 13 units, i.e., between inventory levels 16 and 28.

By construction, the discount factors cause the slopes of the line segments in the benefit function to decrease, thus yielding a concave function that increases at a decreasing rate.

Readiness Drivers Receive Additional Benefit

As previously mentioned, additional benefit was given to readiness drivers as identified by the database of all deadlining demands used by the EDA.[5] So each item was multiplied by a value proportional to the empirically derived probability that the item deadlines an end item. The

---

[5] Peltz et al. (2002).

Figure B.3
Computation of a Benefit Function, Part 1

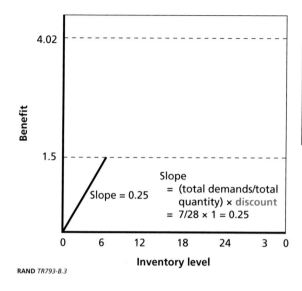

RAND *TR793-B.3*

Figure B.4
Computation of a Benefit Function, Part 2

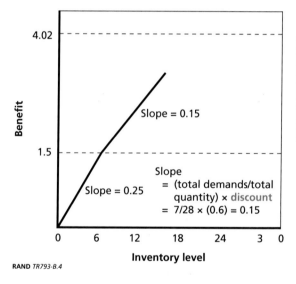

RAND *TR793-B.4*

empirically derived probability was also multiplied by a multiplicative readiness weighting factor chosen by the user.

Specifically, the modified benefit function was derived from the old benefit function as follows:

Figure B.5
Computation of a Benefit Function, Part 3

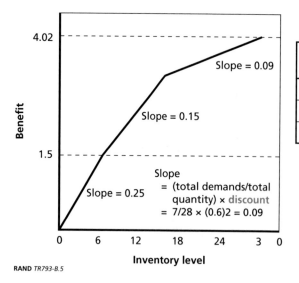

Forward-Stocked NIIN Bucket Table

Bucket Number	Bucket Quantity	Cumulative Bucket Quantity
1	6	6
2	9	15
3	13	28

Endpoint of the third line segment

RAND *TR793-B.5*

$$\text{Modified Benefit Function} = (\text{Old Benefit Function}) \times e^{k \times (\text{criticality})}.$$

The *criticality* of an item is the empirically derived probability that the item deadlines an end item. It is equal to the number of deadlining demands divided by the total number of demands. So zero deadlining demands results in a weighting factor equal to one (e to the zero power equals one). The higher the percentage of deadlining demands for an item, the greater the benefit—which increases the slope of the benefit function. The user chooses the value of the multiplicative readiness weighting factor, k. The larger the value of k chosen by the user, the greater the benefit associated with items with high values of the empirically derived criticality (again, the greater the increase in the slope of the benefit function). So as k is increased, the more likely it is that readiness drivers will be selected for the solution.

Combining Benefit Functions Creates a More Robust Solution

In addition, after the benefit functions were created for each post-offset contingency demand stream, the two benefit functions were combined to create a single concave, piecewise linear benefit function.[6] Although it is possible to give more weight to one contingency demand forecast over another if one contingency demand forecast is considered more likely than another, in this case, we gave equal weight to both contingency demand streams. Also, any number of alternative contingency demand streams can be linearly combined.

Combining multiple benefit functions increases the robustness of the solution by weighting more heavily items and quantities that are common across different contingency demand

---

[6] Note that a weighted average of convex piecewise linear functions is also convex piecewise linear.

forecasts. It also avoids creating a solution that is tailored too closely to one specific instance of contingency demands.[7]

Comparison Across Benefit Functions

Finally, after computing the benefit functions for each item, we compared them across items to find an optimal inventory solution that satisfied the resource constraint. This was achieved through marginal analysis by comparing the benefit-to-cost ratios across NIINs.

The marginal analysis was achieved as described in the following example. Tables B.1 and B.2 show sample marginal benefit values of items at given inventory levels.[8] In the example depicted, there are two items to be compared—NIIN A and NIIN B. Assume that the unit price of NIIN A is $1, whereas the unit price for NIIN B is $10. The third column in each table shows the *benefit-to-cost ratio*, that is, the marginal benefit value divided by the unit price of the item.

Assume that there is a resource constraint of $13. The greatest benefit-to-cost ratio is derived from stocking one unit of NIIN B. Thus, there is $3 left to spend. The second and third greatest benefit-to-cost ratio is derived from stocking two units of NIIN A. After

Table B.1
Benefit Table for NIIN A

Inventory Level	Marginal Benefit	Benefit-to-Cost Ratio
1	0.15	0.15
2	0.15	0.15
3	0.07	0.07
4	0.03	0.03

Table B.2
Benefit Table for NIIN B

Inventory Level	Marginal Benefit	Benefit-to-Cost Ratio
1	2	0.2
2	1	0.1
3	1	0.1

---

[7] Note that, although a combined benefit function will have benefit value *between* the two individual benefit functions used to create the combined benefit function, one should not interpret this to mean that that item has *less* benefit in the combined benefit comparison than that in any individual benefit comparison. Because the benefit comparison across NIINs uses marginal analysis, the relative benefit of a given item is what is used to prioritize the item for war reserve, rather than the absolute benefit value of the item. Specifically, benefit functions from different contingency demand forecasts cannot be compared.

[8] The marginal benefit value is the slope of the line segment in the piecewise linear benefit function.

these purchases, there is $1 left to spend. Although the fourth greatest benefit-to-cost ratio would be for stocking a second unit of NIIN B, there are insufficient funds to purchase another unit of NIIN B. Thus, another unit of NIIN A is purchased, and the resource constraint has been met. The optimal inventory in this case is to stock four units of NIIN A and one unit of NIIN B.

By investing the next dollar on the item with the largest benefit-to-cost ratio, limited funds are optimized to select the items providing the most benefit. This marginal analysis is optimal because of the use of concave benefit functions.[9]

This resource allocation methodology is flexible and fast. By choosing the desired weighting of the time periods and readiness weighting factor, one can rapidly compute and compare different resource allocations, for different resource constraints and for different priorities, with respect to the degree of emphasis on item criticality versus time elapsed in the contingency, thus optimizing WRSI inventory to suit a desired goal.

More generally, if there is more than one resource constraint, a linear programming level approach can be used instead of marginal analysis.

_____

[9] See Everett (1963), Fox (1966), and Sherbrooke (2004).

Bibliography

Amouzegar, Mayhar, Ronald McGarvey, Robert Tripp, Louis Luangkesorn, Thomas Long, and C. Robert Roll, Jr., *Evaluation of Options for Overseas Combat Support Basing*, Santa Monica, Calif.: RAND Corporation, MG-421-AF, 2006. As of December 8, 2009:
http://www.rand.org/pubs/monographs/MG421/

Amouzegar, Mayhar, Robert Tripp, Ronald McGarvey, Edward Chan, and C. Robert Roll, Jr., *Supporting Air and Space Expeditionary Forces: Analysis of Combat Support Basing Options*, Santa Monica, Calif.: RAND Corporation, MG-261-AF, 2004. As of December 8, 2009:
http://www.rand.org/pubs/monographs/MG261/

DoD Directive 3110.6, *War Reserve Materiel Policy,* November 9, 2000. As of January 7, 2008:
http://www.acq.osd.mil/log/sci/policies/DoDD%203110.6.pdf

Everett, Hugh III, "Generalized Lagrange Multiplier Method for Solving Problems of Optimum Allocation of Resources," *Operations Research*, Vol. 11, 1963, pp. 399–417.

Fox, Bennett, "Discrete Optimization via Marginal Analysis," *Management Science*, Vol. 13, No. 3, 1966, pp. 210–216.

Peltz, Eric, Kenneth J. Girardini, Marc Robbins, and Patricia Boren, *Effectively Sustaining Forces Overseas While Minimizing Supply Chain Costs: Targeted Theater Inventory*, Santa Monica, Calif.: RAND Corporation, DB-524-A/DLA, 2008. As of December 8, 2009:
http://www.rand.org/pubs/documented_briefings/DB524/

Peltz, Eric, Marc Robbins, Patricia Boren, and Melvin Wolff, *Diagnosing the Army's Equipment Readiness: The Equipment Downtime Analyzer*, Santa Monica, Calif.: RAND Corporation, MR-1481-A, 2002. As of December 8, 2009:
http://www.rand.org/pubs/monograph_reports/MR1481/

Peltz, Eric, Marc Robbins, Kenneth J. Girardini, and John Halliday, *Sustainment of Army Forces in Operation Iraqi Freedom: Major Findings and Recommendations*, Santa Monica, Calif.: RAND Corporation, MG-342-A, 2005. As of December 8, 2009:
http://www.rand.org/pubs/monographs/MG342/

Sherbrooke, Craig, *Optimal Inventory Modeling of Systems: Multi-Echelon Techniques,* 2nd ed., Norwell, Mass.: Kluwer Academic Publishers, 2004.

Snyder, Don, Patrick Mills, Adam Resnick, and Brent Fulton, *Assessing Capabilities and Risks in Air Force Programming: Framework, Metrics, and Methods*, Santa Monica, Calif.: RAND Corporation, MG-815-AF, 2008. As of December 8, 2009:
http://www.rand.org/pubs/monographs/MG815/